Marine Habitats of Cape Cod

Gil Newton

Published by Gilbert Newton
gdnewton@comcast.net

Copyright 2017 by Gilbert Newton
All rights reserved.
Published in the United States 2017.

ISBN 978-0-9978182-4-6

To purchase a copy of this book, contact:
Gilbert Newton
P.O. Box 2051
Cotuit, Massachusetts, MA

Book produced by Nancy Viall Shoemaker, West Barnstable Press

Table of Contents

Foreword	iii
Diversity of Marine Habitats	1
The Sandy Beach	7
Tide Pools	9
Sand Dunes	11
Salt Ponds	17
Rock Jetties	21
Docks and Piers	25
Salt Marshes	29
Tidal Flats	33
Estuaries	37
An Ocean Plan	39
Outdoor Education and Activities	43
Glossary	50
Recommended References	53
About the Author	54
About the Photographer	55
Acknowledgements	56
Colophon	58

Foreword

by C. Eben Franks

For more than 50 years, Gil Newton has devoted his life to observing, documenting and lecturing on the plants, fish, seaweeds and shellfish indigenous to the coastal regions of Cape Cod. His infectious enthusiasm and highly evolved teaching skills have inspired thousands of students to further their studies of ecology, coastal zone management, seaweeds and tidal processes.

It is with growing concern that Gil and his students are seeing the insidious effects of ocean acidification, nutrient loading, toxins, warming oceans, rising sea-level, invasive species, plastics in the marine environment, shifts in boundary currents, and collapse of fisheries. With these shifts and threats to the ecosystem, Gil's guide books become valuable waypoints that define and describe the interrelationships between development, ecology, agriculture and the 21st Century technology-based society in which we live. It becomes clear that decisions and choices made on scales from global to regional to individual can and do have impacts that affect us all and everything we see.

The best way to get a grasp on the forces that are shaping our environment and the ecology of which we are a part is to spend time in the natural world, looking closely at the ways humans, animals, plants, geology, climate and weather are inextricably intertwined. With Gil Newton's guide books at hand, we find great insights to the complexity and beauty of the world around us.

Inspired and informed by thousands of days hiking, observing, recording, lecturing and collecting samples and data around the Cape, Gil Newton has accumulated a wealth of insights into the landscape and coastal ocean he so dearly loves. His eagerness to share this knowledge and raise awareness have been the guiding principles of his relentless efforts. Gil is our present-day incarnation of the likes of Henry David Thoreau, Louis Agassiz and Henry Beston. He recognizes the multifarious threats to the Cape's ecology and is a leading voice in conservation, education and advocacy for its protection and enlightened use.

There is beauty in Gil's lifelong passion. It derives directly from the Cape and the life within and around its waters.

C. Eben Franks is an ocean explorer, researcher, mariner and educator and has spent more than 8 1/2 years at sea on research expeditions world-wide since 1971. Since retiring in 2014 he devotes most of his time to educational and research projects.

"There is an adventure waiting for anyone who explores the diverse and fascinating marine habitats on Cape Cod."

Gil Newton

Gil Newton photo

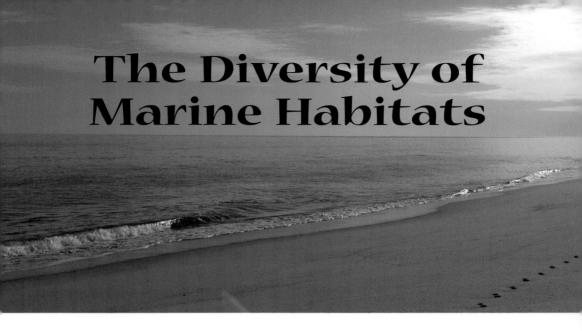

The Diversity of Marine Habitats

One of the most interesting features seen in the study of the coastal environment is the wide diversity of marine habitats. It is vital to examine both the physical and biological characteristics of a habitat to understand the presence and distribution of life forms there. The plants and animals that make their homes in any part of the ocean have special requirements for their survival and sustainability. In fact they often change their own habitats, modifying them in ways that affect themselves and other species.

From sandy beaches to rocky shores to tidal flats to salt marshes, coastal marine habitats have very unique parameters that define them. Some of the most significant and measurable characteristics include the range of temperature, salinity, and dissolved oxygen. These in turn can be affected by the slope of the beach, the particle size of the sediments, and the exposure to waves and currents. In shallow areas such as bogs and estuaries an algae population bloom can have an important impact on dissolved oxygen (DO) levels. The resulting sharp decrease in DO during decomposition causes declines in fish, mollusk, and crab.

Breaking waves can also have a major effect on the distribution of organisms. The weight of the water as it strikes the substrate removes the smaller and lighter particles which can affect the amount of oxygen present. In order for animals to survive this dynamic and shifting environment they need to burrow deep into the ground or locate an object for attachment. Such turbulence along the shore can result in an eroded and constantly changing habitat.

Marine habitats on Cape Cod are also affected by nutrient loading, particularly nitrogen. Sources of nitrogen include septic systems, road runoff, lawn fertilizers, and atmospheric fallout from the burning of fossil fuels. This can lead to accelerated eutrophication of a shallow system and the algae blooms mentioned earlier.

Cape Cod is blessed with its diversity of marine habitats. There are several barrier beach systems in which salt marshes, estuaries, tidal (mud) flats, sandy beaches, rocky shorelines, sand dunes, and salt ponds are all present. Each of these microhabitats contains a specific assemblage of plants and animals interacting with one another and their physical environment. To fully understand the ecology of the shoreline one must carefully examine each of these habitats and discover the full range of species interactions. Organisms are not uniformly distributed throughout the marine environment. They are often located in distinct bands or zones along the shore. Physical factors such as wave action, temperature, salinity, and DO will influence this zonation. It will also vary from place to place. For example, a rocky shoreline may exhibit clearly defined zones characterized by bands of green algae, barnacles, and rockweed.

However, it's a different situation on a sandy beach where the substrate may shift frequently due to wave activity. Most of the species living there are burrowing deep in the sediments to avoid being dislodged.

Many marine species are part of the benthic zone which includes all those that live on or in the bottom or substrate of the ocean. For the coastline this would include the marine plants that need sun for photosynthesis from the shallow water seaweeds to the subtidal eelgrass beds. Most of the animal phyla are represented here as well.

The benthic zone can also be divided into those animals that live on the surface of the substrate (epifauna) to the burrowing species (infauna). Some are found in the dynamic intertidal zone which is the area between high and low tide. Not only does this substrate shift with the waves and tides, but animals here have evolved strategies to survive periodic exposure to the air.

Animals of the benthic environment have also evolved different symbiotic relationships. The carapace of a spider crab (*Libinia emarginata*) may be covered with barnacles, anemones and sponges. These hitchhikers are moved to areas where there may be food available while giving the spider crab added protection as camouflage. This process in which all the species benefit is called mutualism. On the other hand some mollusks may harbor small, harmless worms in their shells. The relationship may benefit the hidden worm, but doesn't help or harm the host. This is called commensalism. However, if the worm obtains nutrition from its live host, thus weakening or harming it, then it's an example of parasitism.

One of the most interesting residents of the infauna is the attractive plumed worm (*Diopatra cuprea*). This polychaete worm lives in a tube that is often covered by bits of shell and sediments glued there by the animal's mucous. Broken tubes can be found along the shore where this animal lives. Infauna worms are also found in the deep parts of the ocean benthic environment where they feed on organic debris that drifts down from the surface. This is a world largely unexplored but may consist of many unnamed species. It's quite possible for these dark and deep sections of the ocean to contain a high diversity and biomass.

Other methods of feeding include the filter feeders such as clams and other bivalves which bring in water through a siphon and sorts small food particles for consumption. Jellyfish use their tentacles to capture their prey immobilizing them with their stinging cells or nematocysts. Some polychaete worms remove food material from ingested sediments. Sea stars pull a mussel's shell apart with their strong arms and then move their stomach into the prey for digestion. Many marine crabs have a very diverse diet feeding directly on both live and dead material.

The predators at the top of the food web are fewer in number and biomass. They have a significant ecological role in the food web. Not only do they prevent the overpopulation of smaller species, but they can increase biodiversity by making more available habitat for colonization by others. For example, mussels can form very dense populations making it difficult for other species to live and grow in that area. A group of sea stars will feed on the mussels and open up additional space for byozoans, anemones, worms, and mollusks. Having strong healthy links at each trophic level is needed to maintain a sustainable biodiversity.

The Sandy Beach

A sandy beach appears to have little life particularly along the water's edge where wave action is continuous. Strong waves remove the finer sediments quickly resulting in coarser material and more rapid drainage. In quiet, protected areas the waves have less energy and the fine sand remains. These conditions are more likely to support animal life, particularly burrowing organisms such as polychaete worms and some clams. Crustaceans such as mole crabs (*Emerita talpoida*) burrow in this environment and filter feed small food particles with their antennae.

The size and composition of beach materials varies with the seasons as well. Sometimes a winter berm will form at the upper portion of the beach. Waves may also remove sand and form a sand bar slightly offshore. In all cases the sandy beach is a dynamic system in which sediments are constantly being moved and relocated. The structure of the sediments may also vary vertically along the shoreline with fine sand particles closer to the water's edge and more gravel type material at the upper section where the wave energy is most intense. The wind can also contribute to the movement of sand and result in the buildup of dunes. These systems of moving sand along beaches and the formation of dunes are called barrier beaches and are very important in protecting upland areas from erosion and coastal storm damage.

Other physical factors along a sandy shoreline include a relatively uniform temperature and salinity profile particularly in areas of fine sand particles. However, oxygen concentrations may vary depending on particle size. The larger the particle of sand, the greater the exchange rate of oxygen between the water and the atmosphere.

Sandy beaches often have a wrack line in which various species of seaweed as well as eel grass wash up on the beach and harbor small animals such as mollusks, sandhoppers, bryozoans, and tube worms. Occasionally a horseshoe crab will swim in the shallow water, and many species of shorebirds can be seen parading up and down the beach edge feeding on mole crabs and other small critters.

Tide Pools

At low tide on some beaches pools of shallow water form that are inhabited by several species of marine organisms. It is amazing how diverse many of these tide pools can be. If they occur regularly as a result of the topography in the intertidal zone then certain species can almost always be found. But these microhabitats frequently change due to wide fluctuations in temperature, oxygen concentration, and salinity. The more shallow the tide pool the more likely it is to heat up rapidly under the warm summer sun and freeze during the cold winter, particularly in areas of freshwater runoff.

Consequently many of the species found in tide pools have a wide tolerance for changing physical conditions. Other environmental factors that affect the presence and distribution of tide pool organisms include wave action, competition, and grazing and predation. The kind of seaweed present depends on the degree of grazing by herbivores such as sea urchins and periwinkles. Green algae such as sea lettuce (*Ulva lactuca*) are more easily grazed than fleshy brown algae like rockweed (*Fucus vesiculosus*).

Animal life may vary as well though tide pools are good places to find sea stars, mussels, hermit crabs, green crabs, periwinkles, shrimp, and sea urchins. If there are numerous rocks present, an abundant barnacle population may be seen. Sea anemones may also attach to the rocks and can feed on small animals washed in by the waves.

If there is a large population of rockweed then many animals can be found attached to the fronds or living underneath the algae. Rockweed retains moisture at low tide and so many species avoid desiccation by hiding beneath its fronds. Small encrusting organisms including bryozoans, hydroids, and tube worms can be seen on the rockweed by using a hand lens.

The general characteristics of rockweed are easy to identify. The brown forked branches have pairs of small air bladders along the midrib of the frond which help the alga float in the water. The alga is attached to a hard surface by a structure called a holdfast. Sometimes the tips are swollen with small bumps which contain the reproductive conceptacles. The male conceptacles release motile sperm cells in the water when the tide comes in. The female conceptacles release the egg cells that are carried by water currents. A single sperm cell fertilizes an egg cell and the resulting zygote settles down on a solid surface. It then grows into a new rockweed plant.

Fucus with tube worms

Tide pools are great habitats to explore and monitor shoreline changes in biodiversity. These dynamic habitats vary considerably from beach to beach in their size, species composition, and physical characteristics.

Sand Dunes

A sand dune forms through the movement of sand particles by the wind. The lighter the sand, the more it moves. The heaviest sand particles settle against an object such as a rock or plant and build the dune around it. Onshore winds transport sand from the beach to a more inland area and this is where the sand begins to accumulate. Without vegetation the sand gradually accumulates and drifts can form that change in size and shape. A sand dune begins to stabilize when several pioneer plant species begin to grow. Over time these plants grow extensive roots which anchor the plant and the leaves start to block the sand from the wind.

A sand dune can move or migrate depending upon the size of the particles and the amount of wind. Dunes are fed by wave action that pile up the sand on the beach. As sand builds up it falls forward and creates a slope. Unless the dune is colonized by a plant community. It will continue to change shape and size.

The most conspicuous plant on a dune is beach grass (*Ammophila breviligulata*) which grows up to three feet tall. The leaves form clumps that trap moving sand, holding the small particles in place. It spreads with underground stems called rhizomes that send up shoots for dune stability.

A low-growing grayish-green shrub called beach heather or poverty grass (*Hudsonia tomentosa*) is also common. This plant is neither a heather nor a grass. It belongs to the rock-rose family, the Cistaceae. Beach heather is an  excellent control agent on dunes. Its deep roots and spreading growth habit help block the removal of sand by wind and rain. It is adapted to this harsh and dry environment and is able to capture and conserve water. From a distance beach heather looks like it grows in dense mats on the sand. A close examination of the leaves reveals a scale-like and alternate arrangement. In early summer it produces an array of small golden flowers that open in the sun. There are five petals and five to 30 stamens. The plant is evergreen and perennial.

Another common dune species is the seaside rose (*Rosa rugosa*) which was introduced to this country in the late 1800's and is often used in landscaping throughout the Cape. Rugosa means wrinkled which describes the surface of the leaf. The plant has a compound leaf with five to nine leaflets. It can withstand

Seaside rose blooming

Seaside rose with its fruit, rose hips

salt spray and sandy soil. The flowers are red, pink or white, and appear in late spring through the fall. The plant grows as a shrub up to five feet tall and produces the fleshy fruits called rose hips which can be made into jelly. In addition to human use the rose hips are an important food source for many animals, particularly when food becomes scarce in the colder months. The seaside rose adds diversity to this assemblage of salt-tolerant, xerophytic (desert-loving) plants. Not only is it a food source for animals, but its dense branches also provide shelter. Many varieties of roses require frequent watering, spraying, and fertilizing. Soil conditions have to be modified and insect pests have to be controlled. On the other hand, *Rosa rugosa* is one member of this huge plant group that has adapted well to the Cape's unique environment.

Dusty miller

Other plant species that contribute to erosion control include dusty miller (*Artemisia stelleriana*), seaside goldenrod (*Solidago sempervirens*), and beach pea (*Lathyrus japonicus*). Like any ecosystem, a dune with a greater diversity of species is more likely to be stable. A monoculture of any plant becomes susceptible to predation and disease.

Goldenrod

Beach pea

Animal life on a sand dune is restricted to those species adapted to harsh conditions, such as extreme temperatures. Camouflaged by the sand and able to adapt to a variety of habitats, the wolf spider (*Hogna* spp.) is a common predator on the dunes. Insects, birds, and small terrestrial mammals often visit and travel through dune systems.

Piping plover, photo by Janet DiMattia

Other species may reside near the toe of dunes such as the threatened piping plover (*Charadrius melodus*) or exist in the interdunal areas such as the Eastern spadefoot toad (*Scaphiopus holbrookii*), also classified as threatened.

Eastern spadefoot toad, photo by Sean Kortis

Sand dune systems are very unstable and fragile. While the plants are salt and wind resistant, they are quite vulnerable to trampling. Where dunes front popular beaches people tend to walk over the plants and create foot paths that eventually erode the area. That is why one of the most common and effective management tools is to construct boardwalks over these sensitive areas.

People prefer to walk on these platforms and they direct foot traffic away from the plants. Regulations exist on the spacing between the boards to allow enough sunlight in for plants to grow under the walkway. In areas already eroded as footpaths, fences can be constructed around them, again framing the area for passage.

Salt Pond Visitors Center, photo by NVS

Salt Ponds

Cape Cod's landscape is characterized by dozens of small kettle ponds formed by the receding glaciers thousands of years ago. In some instances the ocean has broken through creating a salt pond and allowing a salt marsh environment to form a ring around the pond. These ponds are subject to tidal changes and vary in salinity depending on their proximity to the ocean.

Some salt ponds, such as the one at Dowses Beach in Osterville, are connected by a narrow culvert to a bay or estuary. Others, like the salt pond at the Cape Cod National Seashore Salt Pond Visitors Center in Eastham, are part of a large salt marsh system, in this case Nauset Marsh. In both of these examples salt marsh grasses (*Spartina*) support a large diversity of life in and outside the pond. When the grasses break apart and decompose into smaller detritus particles, these provide essential nutrients to many species of animals.

Mollusks are present throughout the system and are able to attach to the muddy substrate or the banks of the salt marsh. The surface of the sediments is populated by an army of mud snails (*Ilyanassa obsoleta*) that are feeding on bits of food.

Mud snail photo by GDN

Ribbed mussels (*Modiolus demissus*) form dense clusters along the edges of the marsh. Quahogs (*Mercenaria mercenaria*) flourish in this environment. There are aquaculture rafts of quahogs growing in the Eastham salt pond. Moon snails (*Lunatia heros*) are common predators on clams, attacking them with their sharp radula.

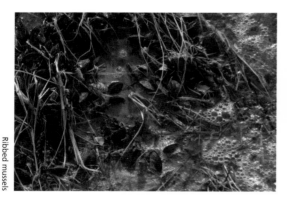
Ribbed mussels

Moon snail

Many crab species can be found in this ecosystem also. Blue crabs (*Callinectes sapidus*) congregate near the entrance to the salt pond at Dowses Beach. Spider crabs (*Libinia emarginata*) favor the muddy substrate as an effective environment for camouflage. And the omnipresent green crab (*Carcinus maenas*) is invasive and a significant predator of commercial shellfish.

Blue crab

Dowses Beach salt pond by NVS

Because many salt ponds are shallow and are located near highways and developed areas, they tend to concentrate nutrients from runoff and other sources. Consequently algae blooms can form where nitrogen loading is an issue. This can have a negative effect on any eel grass (*Zostera marina*) or animal populations. A salt pond can quickly become eutrophied particularly during the warm summer months.

Salt ponds represent a successional change in which a freshwater system becomes marine after a breakthrough or change in the dynamic coastal environment. Fluctuations in the physical environment are accompanied by changes in species structure and diversity.

Rock Jetties

Some of the coastal habitats that are fun to explore include the numerous jetties and groins in most of the towns on the Cape. Several years ago those were built in an effort to control erosion at public beaches and private property. We now know that they have had the opposite effect and actually increase erosion in those areas by focusing wave energy on a narrower section of the beach. Groins are smaller structures than jetties and can be found scattered along the frontal beach. A jetty is located at the mouth of an inlet and is a much larger and longer structure often composed of large boulders.

Still these structures may contain a wide diversity of living organisms that have adapted to the harsh conditions of constant wave, tide, and storm activity. In the jetty at the Cape Cod Canal small pools of water can get trapped at high tide and harbor many marine animals such as sponges, sea anemones and sea stars. I have seen large populations of salps (*Thalia democratica*) get trapped in these pools. Salps are small, transparent planktonic animals that resemble jelly fish and are often seen in chains or large groups. The most conspicuous member on a rock jetty is the common rockweed (*Fucus vesiculosus*), a brown alga that can survive periodic exposure to the air at low tide. Hiding underneath its fronds are bands of small animals such as barnacles, periwinkle snails, and crabs.

However, these algae and animals are not distributed evenly along the jetty. In fact, some of the rocks might be completely devoid of all life, while others have smaller, more randomly distributed populations. This variation may be caused by several physical and biological factors.

The most important physical influence occurs on the side directly facing the waves and tides. Obviously strong wave action will limit the colonization and growth of any critter trying to get established. Even those already present could be removed by strong wave energy from a storm.

Some of the most significant factors are not readily visible. These include the interactions between the different life forms. Because of the limited surface area on the rocks, there is competition for that valuable real estate. Also, some species may only occupy a small portion or band because they are unable to survive prolonged periods of exposure. Some seek shelter under the seaweed fronds to avoid predators or to remain moist.

The rockweed can also influence colonization rates. The movement of the water causes the fronds to sway back and forth, and this motion can brush away any larvae attempting to attach to the rock surface. Other plants and animals may be subject to grazing pressures. Some of these habitats on the Cape contain huge numbers of periwinkle snails which normally graze on small diatoms and algae on the rocks and seaweed. But if their numbers get too high, they may even start to consume the rockweed, affecting its growth and that of many other species. These small variations make this as dynamic a system as that seen along the sandy beach. The zones and sections of living organisms on a jetty are constantly changing as one would expect in the marine environment.

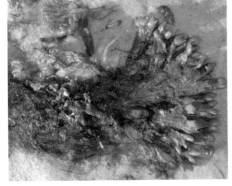

Because of the presence of jetties and groins, some beaches on the Cape have taken on the appearance of a rocky shore. The removal of the fine sand by more intense wave action leaves behind the heavier cobble stones that become places of concealment for animals such as green crabs. Also some seaweed such as rockweed may attach with their strong holdfasts and create new hiding places for small shoreline animals. True rocky shores may exhibit very distinct bands of organisms depending on the section of the intertidal zone. From encrusting algae in the upper portions to the large kelps in the lower end these bands illustrate specific biological and physical requirements for each species.

Pier remnants in Sandwich

Docks and Piers

One of the more interesting marine habitats is found on pilings at docks and piers. Plants and animals exhibit a vertical zonation or distribution on the piling particularly after several years of colonization. There are several species that are frequently found here, but are absent from other habitats.

Several species of algae can be found attached to a piling. The dominant alga is usually sea lettuce (*Ulva lactuca*), consisting of large green sheets that often fold along the edges. Sea lettuce is tolerant to changes in salinity and grows abundantly in areas with high nitrogen runoff.

Marine Habitats of Cape Cod

Another common alga growing on pilings on Cape Cod is Irish moss (*Chondrus crispus*). This has historically been an economically important edible seaweed. Irish moss is a deeply red to purple alga with many flat blades that can measure up to four inches long.

Interspersed among the fronds of the seaweed are several interesting animals. A tunicate called a sea vase (*Ciona intestinalis*) is one of the most common. This vase-shaped filter feeding animal is nearly transparent and grows in large concentrations along the piling. Its body is soft and thin and contains retractable siphons for feeding. This animal also has a notochord, a supportive structure made of cartilage.

Clumps of hydroids called *Tubularia* spp. are also attached to the pilings. Sometimes these animals are mistakenly believed to be algae, but a close look reveals several stem-like structures capped with pinkish polyps. Underwater they appear to be small colonies of pink flowers. However these polyps have tentacles used to capture prey. This group of animals also attaches to the bottom of boats.

Scattered throughout the piling are colonies of rock barnacles (*Semibalanus balanoides*). The larval stage of this crustacean is planktonic. Once it settles on a surface the adult can grow up to one inch long. These animals are also filter feeders and use their tiny feet to capture food particles while waving through the water.

Feeding on the barnacles are several spider crabs (*Libinia emarginata*) that scurry up and down the piling. This crab has a brown carapace and long, spindly legs. Spider crabs have excellent camouflage and are often seen with other animals and algae attached to their shell.

A piling may also harbor other species including sponges, sea anemones, mussels, oyster drills, and bryozoans. This habitat is stationary with a relatively constant physical environment. Over time the entire piling will be covered with an impressive assortment of marine species.

Salt Marshes

Salt marshes form in quiet coastal areas where an incoming tide deposits sediments that are trapped by cordgrass (*Spartina alterniflora*). Usually these areas have reduced erosion pressures and are characterized by a muddy substrate. The salt marsh terrain is quite irregular. Small, shallow and muddy areas form that periodically fill up with water. These pools are called pannes and are often biologically rich with tiny organisms that attract birds and other animals. Many species such as mummichogs and shrimp may get trapped in these pannes. I once found a young one inch horseshoe crab in one.

The pannes tend to be quite salty, particularly in the summer when evaporation is more rapid. I've observed pockets of exposed mudflats around a panne in which the succulent glasswort (*Salicornia europaea*) is the first plant to colonize the area. Later this may be replaced by the salt marsh hay (*Spartina patens*) as succession continues.

Glasswort

Around mid-summer clusters of small purple flowers can be seen in the marsh - it is one of the more popular marsh plants, sea lavender (*Limonium nashii*). Found in the upper marsh amidst salt marsh hay, it can grow between one and two feet tall. When sea lavender is not in flower, it can be identified by a set of flattened leaves on the ground. Unfortunately this plant is frequently collected for dry flower arrangements; this reduces the population by elminating a source of seeds. Please don't pick this plant. Enjoy it in its natural habitat.

There are extensive salt marshes along the east coast of North America, including Massachusetts, although most of them are located in the southeastern United States. The Great Barnstable Marsh consists of about 4,000 acres including the area behind the six-mile Sandy Neck barrier beach. A salt marsh is periodically submerged and contains a significant diverse community of plants and animals. The plants provide a major source of nutrition to large numbers of animals, either through direct consumption or after they have died and decomposed. Such decayed organic material is called detritus and may be transported throughout the marsh and even offshore by daily tidal action. This detritus supports an abundant fauna including clams, mussels, blue crabs, shrimp, and several species of fish.

Salt marshes play an important physical role along the coast. They act as buffer zones between the powerful energy of oceanic waves and the land. In some low lying sections of the country, a salt marsh may help prevent salt water intrusion into drinking water supplies.

The public has now come to understand that salt marshes are not wastelands or simply breeding grounds for greenhead flies. Too many acres have been cleared and filled, thus damaging the coastline by accelerating erosion. In spite of coastal wetland regulations, salt marshes are not fully protected. Incremental losses by encroachment of development will continue to threaten these valuable ecosystems as long as population pressures continue to squeeze the last remaining acres of upland.

Brewster flats, photo by Jim Mills

Tidal Flats

Tidal flats or mud flats are located in quiet sheltered areas on open shorelines where mud and sand are deposited. They are frequently part of a salt marsh system. The muddy substrate has a high concentration of organic detritus populated by large colonies of decomposing bacteria. Dissolved oxygen readings are low particularly in the deeper sediments. The area is exposed at low tide and the substrate may contain several burrowing species as well as a rich meiofauna.

Walking along a mud flat at low tide can reveal the presence of many living things. Some of the common mollusks living in this habitat include the soft shell clam (*Mya arenaria*) which may squirt water from small round holes in the mud. One of the fastest diggers is the razor clam (*Ensis directus*). It uses its foot to dig into the substrate very quickly whenever it needs to hide. The razor clam shell is about five times longer than it is wide and is brown along the edges. These two mollusks are bivalves and filter feeders; they take in water through a siphon to trap and select food particles.

Other animals in this habitat are deposit feeders and include several polychaete worms. These animals obtain their nutrition by consuming the organic material in the sediments. The most common is the clam or sand worm (*Nereis virens*) which can grow over a foot in length. This brightly colored species can bite so caution should be exercised when handling them. Another common worm is the lug worm (*Arenicola marina*) that lives in a U-shaped burrow. Though this animal is inconspicuous along the shore, its castings are commonly found.

Other animals that can be seen include small dark snails that may be feeding on the carcass of a fish or a crab. These scavengers are the eastern mud snails (*Ilyanassa obsoleta*) that move in large numbers to a food source. A crab species included in this area is the blue crab (*Callinectes sapida*), characterized by its blue carapace, saddle-shaped hind legs, and sharp spines on the edges of its shell. The spider crab (*Libinia emarginata*) has a brown spiny carapace that camouflages it against the muddy bottom.

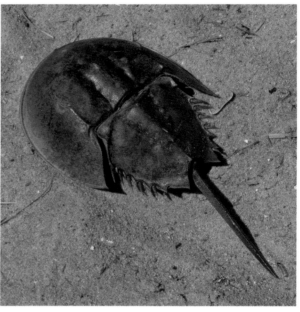

One animal that may be found in both sandy and muddy shores is the horseshoe crab (*Limulus polyphemus*). It has survived the reign of the dinosaurs and continues today, 350 million years after it evolved. Its blue blood contains a substance capable of detecting bacterial endotoxins and a shell with material that can increase the healing of wounds. It looks formidable but is harmless to humans. Indeed it has probably saved many human lives because of its unique medical applications.

Horseshoe crabs are not true crabs but are more closely related to spiders. They have five pairs of legs that enable them to crawl along the substrate where it feeds on worms and mollusks. There is a pair of small pincers called chelicerae which assist the animal in feeding and a set of book gills used for breathing and swimming.

The shell or carapace is actually an exoskeleton. As the animal grows larger it needs to molt. It does this by splitting the carapace and emerging head first. True crabs molt from the back. It's not unusual to find large numbers of recently molted shells on the Cape's beaches in late summer. It takes about twelve hours for the soft new shell to harden.

A female horseshoe crab will lay thousands of eggs during spring full moon high tides. These are then fertilized by the smaller male and soon hatch out. Many migratory birds such as red knots and sanderlings depend on this food source as they fly to northern grounds.

Mud flats are interesting habitats and best observed at low tide, particularly at locations such as First Encounter Beach in Eastham and Crocker Neck in Cotuit.

GDN

Estuaries

One of the most productive ecosystems on Cape Cod is an estuary. This transition system contains several distinct habitats that function interdependently and connect freshwater systems (rivers, groundwater) to the ocean. Consequently salinity varies widely, depending on the topography of the bay as well as the amount of freshwater influence. An estuary is often encircled by salt marsh habitat with sections that are brackish.

Estuaries are very productive systems and support a large diversity of wildlife. Nutrients that support plants and algae, particularly phytoplankton, are abundant in estuaries. This in turn supports many commercially important shellfish and finfish populations. Unfortunately sometimes these nutrients, notably nitrogen, occur in excess and eutrophication conditions can occur. Algae blooms, followed by a reduction in shellfish and crab populations, are the result.

Estuaries are important areas for protecting uplands from coastal storm damage, sheltering migratory birds, and preventing the intrusion of contaminants into the water. Sometimes these systems are well mixed between the salt water and the freshwater input. Freshwater has a lower density than salt water and may perch on its surface.

Many animals that survive in estuaries must have the ability to tolerate changes in salinity (euryhaline species). For example, blueback herring (*Alosa aestivalis*) are anadramous species that migrate from the ocean to freshwater. An animal must have the ability to regulate its water and salt balance if it is to survive under these conditions. However most species are stenohaline and are not adapted to widely fluctuating salinities.

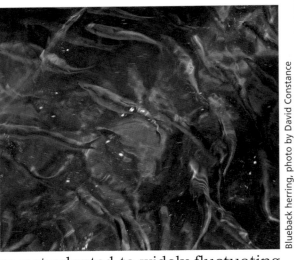
Blueback herring, photo by David Constance

Estuaries are popular areas for recreation. Human pressures from activities such as boating can place environmental stress on the system. Because of their popularity the areas surrounding an estuary are often heavily built up so that effluent from septic systems and runoff from roads can create nutrient loading issues.

An estuary is an excellent example of a rich and diverse ecosystem that includes salt marsh habitat and is influenced by surrounding development. Fishing and boating are popular activities, even sailboarding as seen here at Eastham's First Encounter Beach. Estuaries can also offer the public many educational opportunities.

An Ocean Plan

Today we need to have a strong commitment to protect the ocean. Human activities throughout the planet are having an adverse impact on this most important ecosystem. The ocean covers about 71% of the planet's surface. Municipal and industrial contaminants flow daily from land to the sensitive bays and estuaries around the world. Major fisheries on the continental shelves are severely depleted. Coastal wetlands are under assault from the effects of runoff and development. These are areas that provide nutrients and shelter to a majority of commercial species.

Climate change is raising the temperature of the sea surface. Polar ice is melting, causing the sea level to rise which has led to more rapid coastal erosion. The acidification of oceanic water is affecting numerous species of marine life including coral, mollusks, and echinoderms. Climate change is changing both coastal and oceanic marine habitats.

Meanwhile we continue to exploit marine waters for transportation, recreation, aquaculture, and energy development. Coastal states such as Massachusetts have formed management agencies that address local problems as they occur. But these efforts fall short of what is truly needed.

There must be an international agreement and understanding of the importance of the world ocean. And this can be accomplished by a renewed dedication to its preservation. Similar to the decision to land a man on the moon, a national goal that focuses on marine research and conservation will inspire other nations to do the same.

A management scheme that looks at the entire ecosystem factors in links between the natural features and human influences. Economic and cultural pressures affect many species, especially those used in industry. In the marine environment several economically significant areas interact with each other. These include interests in fisheries, energy development, tourism, and transportation. To balance all of these areas that are sometimes in conflict can be a formidable task.

For example whenever population data suggests that a species might be threatened or endangered, measures are often taken to reverse this decline. But it is inevitable that these steps will impact some other use of the resource. A change in international shipping lanes has been initiated to reduce collisions with right whales. Sections of beaches are fenced off to prevent the disturbance of piping plover and tern nesting sites. Shellfishing is closed in areas subjected to red tide. And sections of the offshore environment may be temporarily closed to ground fishing to allow the recovery of a species.

In each case human activities can continue but may need restrictions and regulations to protect the threatened resource. This is a dynamic situation that requires constant review and monitoring. These restrictions can be lifted once the danger has passed. In all cases such decisions should be based on the best scientific data available.

Hence the need for comprehensive long-term regional plans that take into account the entire ecosystem and not just one small piece. The benefits of such planning allow economic development to proceed while simultaneously protecting environmental resources. In short we need to combine better planning with a new kind of ethic towards the ocean and the environment in general. Developing this ethical view has been an ongoing struggle for decades. Just getting people to use fewer materials such as plastics has not been easy. When we use less there is less to dispose and less entering marine waters. When we precycle we don't need to recycle as much. When we consciously and deliberately reduce our consumption of material goods, we help prevent additional damage to sensitive resources. It seems difficult for many to see the connections between their consumption patterns and the protection of the marine environment.

We have so much more to explore, discover and learn. New international partnerships and increased research will have extensive and environmental benefits for all. Everyone on this planet is dependent on the natural processes in the ocean. It is time to make the protection of the ocean a global priority.

Outdoor Education and Some Activities

Much has been written and said about the disconnect between children and the outdoors these days. Many of us remember that our childhood playtime was electronics-free and involved discovery outside. Nothing can substitute for direct observation and experience in nature when it comes to understanding how living things interact with the physical environment and each other. To fully appreciate the natural world one must use all of his or her senses. A child learns that the environment is fragile and diverse. A few minutes watching an osprey fishing for its young in an estuary or a bee pollinating a group of flowers on a dune is one of the best ways to develop a caring and nurturing attitude towards the natural world. These young observers will one day be given the task of stewardship of this world. They need to learn how to cherish, protect and sustain these critical resources.

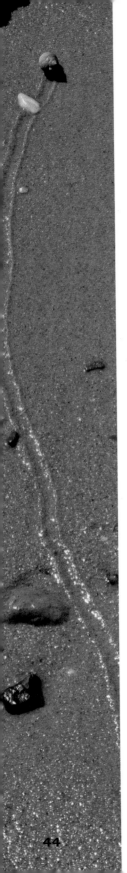

KEEPING RECORDS

One of the most valuable ways of learning about the environment is to keep a journal or log of observations and sketches. This is particularly beneficial over a long period of time in which changes in the habitat can be recorded as well as notations on specific plants and animals. A notebook could be simply a blank sheet of paper or a more detailed data sheet in which specific questions are answered.

ACTIVITY

1) Identify an undisturbed area that can be observed over time.

2) Before you use a field guide to identify an unknown plant or animal, make your own set of observations. This is more effective than thumbing through a book looking at pictures.

3) If it's a plant, try to observe and record information about it when it first appears. Then follow its growth throughout the season.

4) For animals use a hand lens to look for specific features such as the shape of a claw, the size of a shell, or the presence of other attached species.

5) Are there variations within the population of the same species?

6) Are there signs of feeding? For example do some shells exhibit small holes suggesting the work of a predator?

7) Use your observations and a field guide to identify each species and keep an ongoing list.

8) Does the assemblage of species change with different weather conditions or tides?

THE PLOT THICKENS

Different groups and communities of marine species can be identified and compared by selecting a study area along a marine system such as a sandy beach, sand dune, or salt marsh. Comparing one sample plot to another can show changes that have taken place, a phenomenon called succession. Usually these plots are measured in units of square meters but any size area can be studied.

ACTIVITY

1) Select an area that seems to have several species present.

2) Measure and mark off the area with string or ribbon.

3) Count the number of different plants and animals in the study plot.

4) Next count the number of individuals for each species.

5) If it's a common plant like beach grass estimate what percent it covers the plot.

6) On a piece of paper draw a square and label one meter on each side.

7) Measure out a one meter square in the plot.

8) Create a set of symbols, a different one for each animal and plant.

9) On the paper write in the symbol for each species you see.

10) How are the species distributed in the plot? Would you describe them as uniform, random, or clumped?

11) Now count the number of individuals for each species. This tells you the density of the plant or animal per meter squared. (Density = the number of individuals divided by the area or D=N/A.)

12) Do another plot and make the same calculations.

13) How do the two areas differ?

14) What environmental conditions allow some organisms to exist in the plot and not others? For example does the substrate differ? Is wave action stronger in one section over another?

Marine Habitats of Cape Cod

DUST ON THE DUNES

Dusty miller (*Artemisia stelleriana*) is a pale green perennial plant that grows primarily on dunes and sandy soil. It is also a favorite garden plant in sandy habitats like Cape Cod. The leaves are covered with small white hairs and are round at the tips. They grow close to the ground and function effectively in preventing erosion from wind and storms. Flowers are small and yellow on a tall stalk and are wind-pollinated. The plants are grown in gardens primarily for their foliage. When examining any plants on a sand dune caution should be exercised to avoid trampling the plants. Check for tiny deer ticks, a dune being one of their primary habitats. Dusty miller can often be found growing along the edge of dunes so check there first for the plants.

ACTIVITY

1) Rub your fingers along one of the leaves. How would you describe its texture?

2) Do you detect numerous hairs? What do you think is their function?

3) Some observers compare the surface of a sand dune to that of a desert. How are these two environments similar?

4) How does a plant like dusty miller survive in this harsh environment?

5) Are there any flowers present? If so, how would you describe them?

6) How would you describe the distribution of dusty miller on the dune?

7) Are there other plant species growing nearby?

8) If so, how are they similar? How are they different?

9) Why is it an advantage for dusty miller to grow close to the ground?

SAVE THE SEAWEEDS

(From *Seaweeds of Cape Cod Shores* by Gilbert Newton)

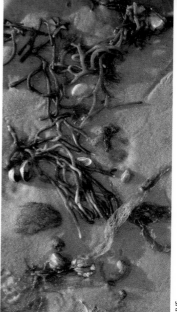

Many of the common seaweeds can be preserved for future reference and study. This method can also be used for various arts and crafts projects using seaweeds; the materials needed are quite simple. The easiest ones to preserve are the stringy, flat, or filamentous forms. The large, thick ones require more complex methods for preservation.

ACTIVITY

1) Visit the beach and look for thin, filamentous seaweeds that have washed up or are present at the water's edge.

2) Make sure that the plants still retain their color. Algae will bleach out their pigments if exposed to the sun for a long time after being stranded on the beach.

3) Collect several of the plants in a bucket and fill the bucket half-way with sea water.

4) Back in the classroom, or at home, obtain a tray or pan large enough for the plant to spread out.

5) Fill the tray with sea water and float the plant in it.

6) Use a thick sheet of paper, such as official herbarium stock, oak tag, or even a large index card.

7) Place the paper in the water and under the plant.

8) Slowly raise the paper so that the plant is picked up.

9) Carefully drain the excess water from the paper.

10) Using a medicine dropper, squirt water around the seaweed to arrange it the way you want to see and remove small pieces and debris that may remain.

11) Cover the plant with cheese cloth or wax paper.

12) Place between several sheets of newspapers, and place several large books on it. (Note: Small plant presses can also be purchased for this stage as well.)

13) The newspapers should be changed the next day because they absorb the water.

14) In 2-3 days, carefully peel off the cheese cloth or wax paper, and the seaweed will remain stuck to the herbarium paper.

15) You should label your specimen with information that includes its name, habitat, and date of collection.

PLANTS IN WINTER

Winter is a very harsh time of year for plants in the northeast, particularly in habitats such as sand dunes. While most parts of the plant die back in winter, several others have adapted to the heavy snows and cold winds. One of the most important adaptations is the prevention of water loss and water freezing. Some of the evergreen shrubs will droop their leaves to protect the living tissues inside. Many of the dune plant species retain the above ground part to protect the sand from eroding away in harsh weather.

ACTIVITY

1) Examine the characteristics of a plant on a dune in winter.

2) Are there any leaves still on the plant?

3) If so, is this plant an evergreen or are the leaves brown and still clinging to the branches?

4) Do the leaves fold downward? If that is the case, how does that help the plant survive in the winter?

5) If there are no leaves on the plant, look for the place where they were once attached. These are known as leaf scars.

6) Can you tell from the leaf scars if the leaves were alternate, opposite, or whorled?

7) Are there any dried fruits, pods, or seeds? Look for rose hips for example. If present, describe their shape.

8) If all the seeds are gone, what do you think happened to them?

9) Some plants in late winter begin to show buds. Can you locate any of these?

10) Based on your observations can you determine if it was insect pollinated or wind pollinated? What would you look for?

Glossary

AIR BLADDERS – gas-filled structures found in rockweed which help the plant float to maximize photosynthesis.

ANADRAMOUS – an animal that moves from the ocean to freshwater in order to spawn.

BENTHIC ZONE – the substrate area of the ocean.

BIODIVERSITY – the variety of life in an ecosystem such as the ocean.

CARAPACE – the "shell" of a crab.

CHELICERAE – a small pair of pincers which assist a horseshoe crab in feeding.

COMMENSALISM – a symbiotic relationship between two species in which one benefits and the other neither benefits nor is harmed.

CONCEPTACLES – the reproductive structures in rockweed that release the gametes into the water.

DENSITY – the number of individuals divided by the area (length x width).

DESICCATION – the loss of water in an organism after exposure to the air and sun.

DETRITUS – decomposing organic material that provides a food source for animals.

ENDOTOXIN – a poisonous substance found in the walls of some bacteria.

EPIFAUNA – animals that live on the surface of the substrate.

EURYHALINE – animals that have the ability to tolerate changes in salinity.

EXOSKELETON – an external skeleton of an animal.

FILAMENTOUS ALGAE – a group of algae that are thin and thread-like.

FILTER-FEEDER – an animal that obtains its nutrition by straining food substances from water.

FROND – the leaf-like structure in algae.

FOOD CHAIN – the transfer of energy from one species to another through consumption.

FOOD WEB – the interaction of several food chains.

GROINS – small rock structures found scattered along the frontal beach.

HABITAT – the place where an organism obtains all of its requirements for life.

HOLDFAST – a root-like structure in seaweeds that attaches to a substrate.

INFAUNA – animals that burrow in the substrate.

INTERTIDAL ZONE – the area along a shoreline between high and low tides.

JETTY – a structure made of rocks built at an angle along the coastline in order to block the movement of sand.

KETTLE POND – a shallow body of water formed by retreating glaciers.

MEIOFAUNA – small animals that live in the sediments along the shore.

MIDRIB – a supportive structure that runs along the center of a rockweed alga.

MONOCULTURE – a single species growing in a designated area.

NUTRIENTS – inorganic substances such as nitrogen that are used by plants for growth.

MUTUALISM – a symbiotic relationship between two species in which they both benefit.

PANNES – pools of water in a marsh that provide habitat for many small animals.

PLANKTONIC ZONE – an area of the ocean characterized by small drifting plants and animals.

RADULA – a small toothed structure in snails used for feeding.

SALINITY – a measure of the salt concentration in water usually in parts per thousand.

SCAVENGER – an animal that feeds on dead animal matter.

STENOHALINE – animals not adapted to widely fluctuating salinities.

SUBSTRATE – a place or object on which an organism is attached or lives.

XEROPHYTE – a plant that can survive in very dry conditions.

ZONE – a division of the shoreline characterized by certain plants and animals.

ZYGOTE – the cell product of fertilization.

Recommended References

Bertness, Mark D. ***The Ecology of Atlantic Shorelines***. Sinauer Associates, Inc. Sunderland, MA. 1999.

Carson, Rachel. ***The Edge of the Sea***. Houghton Mifflin Company. Boston, MA. 1955.

Hay, John and Peter Farb. ***The Atlantic Shore***. Harper Colophon Books. New York and Evanston. 1966.

Newton, Gilbert. ***Discovering the Cape Cod Shoreline***. West Barnstable Press, MA. 2012.

Newton, Gilbert. ***Coastal Corners of Cape Cod***. West Barnstable Press, MA. 2015.

Sterling, Dorothy. ***The Outer Lands***. The American Museum of Natural History, New York. 1967.

Teal, John and Mildred. ***Life and Death of the Salt Marsh***. Ballantine Books, Inc. New York, NY. 1969.

Waller, Geoffrey, Ed. ***Sealife - A Complete Guide to the Marine Environment***. Smithsonian Institution Press. Washington, D.C. 1996.

The Author

Gilbert Newton is a Cape Cod native who has been teaching marine and environmental science at the Cape Cod Community College and Sandwich High School for many years. His classes include coastal ecology, botany, coastal zone management, general ecology, and environmental technology. In 2013 he was appointed the first Director of the Sandwich STEM Academy. He completed his graduate work in biology at Florida State University. He is also the Program Director of the Advanced Studies and Leadership Program at the Massachusetts Maritime Academy and sponsored by the Cape Cod Collaborative. Gil is one of the founders of the Barnstable Land Trust, a former chairman of the Barnstable Conservation Commission, and past president of the Thornton W. Burgess Society and of the Association to Preserve Cape Cod. He is the author of *Seaweeds of Cape Cod Shores, The Ecology of a Cape Cod Salt Marsh, Discovering the Cape Cod Shoreline,* and *Coastal Corners of Cape Cod.*

Photo by Libby Smith

The Photographer

Chris Dumas has lived and worked on Cape Cod for many years. He teaches earth and space science at Sandwich High School and is an advisor to the photography club there. Chris has been involved with outdoor education for most of his career. Photography has been an important part of Chris' life for the last decade. He has traveled around the country in search of interesting vistas. Chris has a graduate degree in Resource Conservation from the University of Montana and is a native of the St. Lawrence River region of New York. His photography can also be seen in *The Ecology of a Cape Cod Salt Marsh, Discovering the Cape Cod Shoreline,* and *Coastal Corners of Cape Cod.*

Acknowledgements

I am very grateful to the following individuals for their assistance and support:

To **David Constance**, **Janet DiMattia**, **Sean Kortis**, **Jim Mills**, **Libby Smith** and **Nancy Viall Shoemaker** for their generous donations of outstanding photographs.

To **Chris Dumas** whose excellent photography and illustrations enhance the stories in this book.

To **C. Eben Franks** for his wonderful foreword and his tireless commitment to science education and research on Cape Cod.

To **Nancy Viall Shoemaker** for her continuing professional advice and guidance in the design and creation of all my books.

This book was designed and typeset by Nancy Viall Shoemaker of West Barnstable Press, West Barnstable, Massachusetts - www.westbarnstablepress.com. The text font is Bookman Old Style, designed by Alexander Phemister in 1858 - working from the font Caslon. It was constructed with straighter serifs, allowing it to keep its readability at small sizes. Photo credits were set in Frutiger, designed by Swiss typographer Adrian Frutiger (1928-2015). The font used for the chapter heads is Nueva which was designed by Carol Twombly for Adobe in 1994. *Marine Habitats of Cape Cod* was printed on 100 lb. white matte stock with a 12 pt. laminated cover.

Printed on recycled paper

"The best way to get a grasp
on the forces that are shaping
our environment and the ecology
of which we are a part is
to spend time in the natural world."

C. Eben Franks